SAE EDGE™
RESEARCH REPORT

Unsettled Technology Domains in Industrial Smart Assembly Tools Supporting Industry 4.0

Thorsten Roye, Dipl.-Ing., M.Sc.
Airbus Operations GmbH

EDGE DEVELOPMENT TEAM

Lars Burkhardt, *Atlas Copco*
Baptiste Bertrand, *Airbus Operations SAS*
Pierre Deffrennes, *Desoutter Tools*
Florian Hanke, *HS-Technik*
Thomas Schönbeck, *Airbus Operations GmbH*
Sebastian Ebeling, *Siemens Gamesa*

SAE INTERNATIONAL®
Warrendale, Pennsylvania, USA

About the Publisher

SAE International® is a global association of more than 128,000 engineers and related technical experts in the aerospace, automotive, and commercial-vehicle industries. Our core competencies are life-long learning and voluntary consensus standards development. Visit sae.org

SAE EDGE™ Research Report Disclaimer

SAE EDGE Research Reports focus on topics that are dynamic, in which knowledge is incomplete, and which have yet to be standardized. They represent the collective wisdom of a group of experts and serve as a practical guide to the reader in understanding unsettled subject matter. They are not meant to provide a recommended practice or protocol. The experts have assembled as a community of practitioners to contribute and collectivize their thoughts and points of view; these are not the positions of the institutions or businesses with which they are affiliated, nor is one contributor's perspective advanced over other contributors. SAE EDGE Research Reports are the property of SAE International and SAE alone is responsible for their content.

About This Publication

SAE EDGE Research Reports provide state-of-the-art and state-of-industry examinations of the most significant topics in mobility engineering. SAE EDGE contributors are experts from research, academia, and industry who have come together to explore and define the most critical advancements, challenges, and future direction in areas such as vehicle automation, unmanned aircraft, cybersecurity, advanced propulsion, advanced manufacturing, Internet of Things, and connectivity.

Related Resources

SAE MOBILUS® Advanced Manufacturing Knowledge Hub
https://saemobilus.sae.org/advanced-manufacturing/

SAE Team

Frank Menchaca, Chief Growth Officer
Michael Thompson, Director of Standards, Information and Research Publications
Monica Nogueira, Director of Content Acquisition
Beth Ellen Dibeler, Product Manager
William Kucinski, Managing Technical Editor

Printed in USA

EPR2020018
ISSN 2640-3536
e-ISSN 2640-3544
ISBN 978-1-4686-0242-5

To purchase bulk quantities, please contact: SAE Customer Service

E-mail: CustomerService@sae.org
Phone: 877-606-7323 (*inside USA and Canada*)
+1-724-776-4970 (*outside USA*)
Fax: +1-724-776-0790

https://www.sae.org/publications/edge-research-reports

About the Editor

Thorsten Roye is the Manufacturing Engineering Expert for Fastening Systems at Airbus. In this role, he serves as an advisor to management and all production plants across Airbus for final and temporary fasteners, fastening tools, and fastening processes for commercial aircraft, defense and space, and helicopters.

He has a diploma in mechanical engineering–aircraft systems (Technical University in Hamburg, Germany) and a Master of Science in Composites (Private University Hannover, Germany).

Mr. Roye has worked for Airbus for the past 15 years, having been responsible for automation, robotics, and mechatronics. Currently, he leads Airbus collaboration with multifunctional teams in procurement, engineering, quality, and research and development. He represents Airbus in several workshops, training programs, and conferences worldwide.

Named on more than five issued patents, Mr. Roye is highly interested in assembly and manufacturing topics touching a broad range of industries, including automotive.

He is an assistant professor at two German universities and focuses on assembly technologies. He is also an instructor for SAE International teaching "Introduction to Airframe Engineering—Design for Manufacturing, Assembly and Automation."

contents

Atosan/Shutterstock.com

Unsettled Technology Domains in Industrial Smart Assembly Tools Supporting Industry 4.0

Abstract

This SAE EDGE Research Report identifies key unsettled definitions of the role of smart assembly tools in the Industry 4.0 world. "Smart" refers to tools that are "specific, measurable, achievable, reasonable/realistic, and time bound." Smart assembly tools are used in all industries, including automotive, aerospace, and space. These tools are employed for measuring, inspecting, gauging, drilling, and installing all existing fastening systems. The role of an assembly tool inside the Industry 4.0 environment is quite important as the smart, intelligent assembly tools have an enablement function. Smart assembly tools have a huge influence on Information and Communication Technology (ICT), assembly cost reduction, process control, and even the product and process quality.

These four four domains—and their undefined nature—are the focus of this SAE EDGE Research Report. The technical issues identified here need to be discussed, the goals clarifying the scope of the industry-wide need to be aligned, and the issues requiring standardization need prioritized.

NOTE: SAE EDGE Research Reports are intended to identify and illuminate key issues in emerging, but still unsettled, technologies of interest to the mobility industry. The goal of SAE EDGE Research Reports is to stimulate discussion and work in the hope of promoting and speeding resolution of identified issues. SAE EDGE Research Reports are not intended to resolve the challenges they identify or close any topic to further scrutiny.

THORSTEN ROYE, DIPL.-ING., M.SC.
Airbus Operations GmbH

Edge Development Team

Lars Burkhardt, *Atlas Copco*
Baptiste Bertrand, *Airbus Operations SAS*
Pierre Deffrennes, *Desoutter Tools*
Florian Hanke, *HS-Technik*
Thomas Schönbeck, *Airbus Operations GmbH*
Sebastian Ebeling, *Siemens Gamesa*

ISSN 2640-3536

Introduction

State of the Industry

The previously published SAE International journal article "Smart Portable Tools: The Practical Solution to Agile Assembly, 2013-01-2295" explains the needs of smart tools at Boeing production sites (with a strong focus on semiautomated drilling tools from Novator). It describes what these special orbital drilling tools can do for the production of aircraft, and it also briefly discusses monitoring features and quality traceability. This topic received updated coverage in the SAE International technical paper "Improving Manufacturing Efficiencies through Industry 4.0 Technologies in Aerospace."

These two publications describe special tools in the aerospace sector—those that are used globally in most advanced industries for measuring, inspecting, and gauging operations, as well as for drilling and fastening processes. We could say that Industry 4.0 is the route for companies to implement manufacturing automation for productivity, reliability, innovation, and repeatability (Figure 1).

Yet, due to the increased complexity of mobility products, the required weight and cost savings, and the high ramp-up and production rates, assembly processes need semiautomated processes for flexibility. In semiautomated assembly processes, only the main process is fully automated—operators must supplement the process with handheld tools. An example might involve hand-tightening a fastener to a specific torque, which is monitored and controlled by a torque sensor to end the process, capture data, and provide feedback to the operator as to the quality of the installation. Assembly processes like this can be found in all industries for different activities (Figure 2).

These assembly processes are quite different, but they all share a commonality regarding smart functions. "Smart," a trendy word for "intelligent," is mostly linked to the Industry 4.0 context. In the case of this report, smart (or "SMART") can serve as an acronym for "specific, measurable, achievable, reasonable/realistic and time bound." The smart approach comprises many features and technologies, but also "wishes and dreams." This SAE EDGE Research Report considers the following four areas to be unsettled domains in definition and standardization, as marked by an apparent lack of common understanding regarding key aspects of each:

- Information and Communication Technology (ICT)
- Cost reduction
- Process Control
- Quality

Figures 3 and 4 depict the backbone and pillars of the four abovementioned unsettled domains:

- Company Strategy
- Company Culture
- Information Technology (IT) infrastructure
- Product Data

All these key aspects are discussed in this report.

NicoElNino/Shutterstock.com

FIGURE 1. **Productivity, reliability, innovation, and repeatability can all greatly benefit from process automation.**

FIGURE 2. **Measuring, inspecting, gauging, lock bolting, riveting, blind bolting, tightening, and drilling are assembly processes found in all types of manufacturing industries.**

Mr.1/Shutterstock.com

FIGURE 3. **The four unsettled domains of smart assembly tools.**

© SAE International.

FIGURE 4. The backbone and pillars of smart tooling.

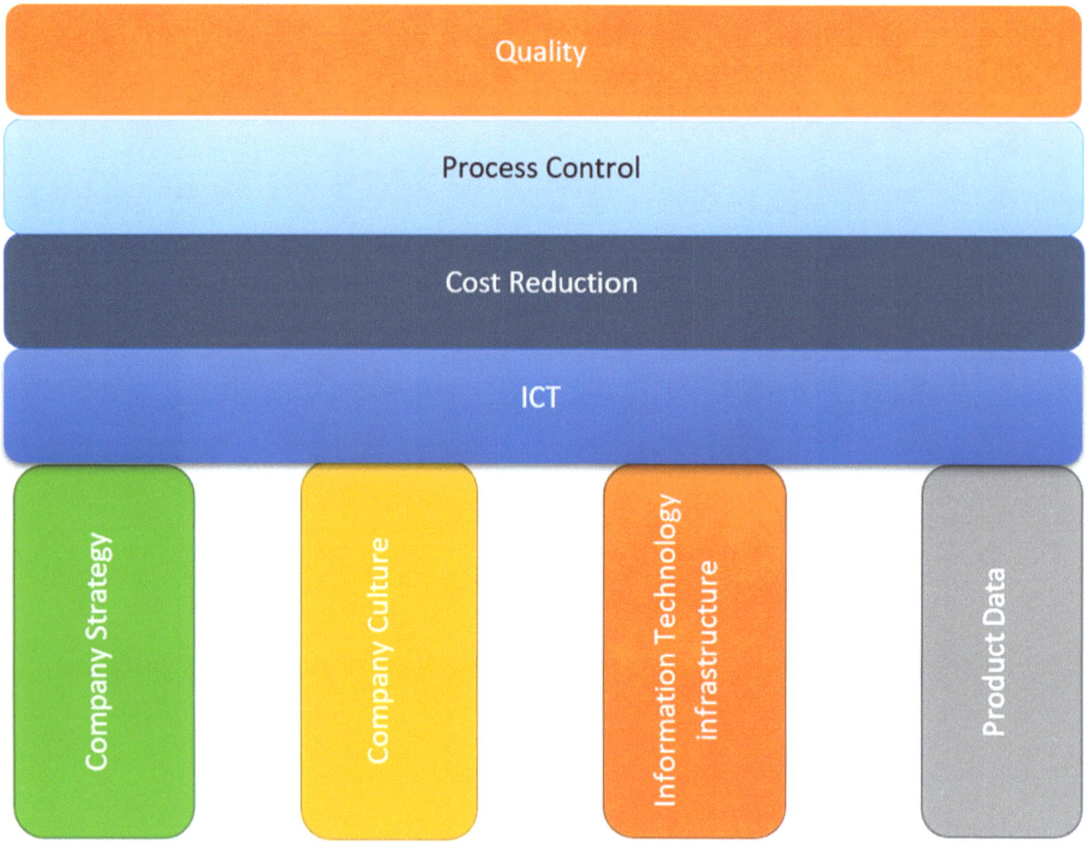

Tooling manufacturers are the most active group in this field to date. All tooling manufactures are designing and developing assembly tools to meet the requirements from a range of different companies from a range of different industries. Therefore, different understandings and interpretations of the needs of product manufacturers currently exists. Additionally, the different tooling criteria—which are driven by different industries—are quite inhomogeneous and consequently the design of smart assembly tool features vary from one tooling manufacturer to another. These main tooling manufacturers which cater to a range of industries include Atlas Copco, Desoutter Industrial Tools, HS-Technik, Gage Bilt, AMDP, Apex Tool Group, Lübbering, Amanco Tool, Arconic Fastening Systems, LISI Aerospace, Bosch/Rexroth, and many others.

With such a wide range of industries and tooling manufacturers, there is the need for a common understanding and standardization of smart assembly tool features. The benefits of this harmonization are twofold: it enables the integration of different brands in one enterprise system and reduces the work involved when comparing various technologies.

Information and Communication Technology

Digitalization

There are many integral concepts of the "smart" industry. Industry 4.0 hinges on automation, big data, system flexibility, and intelligent user interfaces, to name a few (Figure 5). While each preceding industrial shift (i.e., Industry 1.0, 2.0, and 3.0) featured their own identifying technological revolution, Industry 4.0 leverages digitalization. In a digital world, data and knowhow are key factors of company success and competitiveness. Simply put, information and communication technology (ICT) are integral to Industry 4.0.

Smart assembly tools must therefore also fit into this digital world, which means sensors and electronics must be onboard to enable data collection, storage, and analysis.

FIGURE 5. **Industry 4.0 relies on the integration of many complex systems.**

elenabsl/Shutterstock.com

Especially in an assembly production line, everything starts and fails with the operator. Operators are working with assembly tools and these tools must support them to do their job as required by the product or process specification. Currently—in both the aerospace and automotive industries—operators are guided mainly by paperwork specifications to fulfill different tasks during a shift or manufacturing routine. In complex areas, this can be hundreds of different small jobs (e.g., take a specific bolt and nut, use the right tooling and program, etc.). Smart assembly tools could decrease the amount of specification referencing and guide operators with the goal of lower failure rates and accelerated manufacturing processes.

In a rapidly changing world, the processes and products are also changing quickly. For example, engineering requirements are improving due to lessons learnt from in-service behavior, which could lead to optimization of torque and speed values. Other changes could be the result of customized products, where the customer might select specific product features late in the assembly process—this could change many things and creates the need for new operator guidance. Smart, connected tools can be programmed to the right parameters in little time to guide the operator through the new manufacturing process.

Throughout this, the operator also needs physical freedom to move and to do their job, meaning electrical tools equipped with battery power should be the baseline. Wireless tools are key for an autonomous and flexible assembly line. However, besides all positive digital aspects, the cyber security shall not be neglected.

Wireless Connectivity

On top of the electrical power, smart assembly tools must store and analyze data. As discussed previously, wireless capability is key. Dedicated and special systems shall be used, such as Wi-Fi, Bluetooth, or any other kind of wireless protocol. Security levels must be therefore be respected—an example of adequate security measures would be the use of a dedicated Wi-Fi MHz channel with the appropriate level of encryption (i.e., Wi-Fi 6 with Wi-Fi Protected Access (WPA)).

Standardized Interface/ Protocol

For a complex product or process, complex assembly tools are needed. Because every tooling manufacturer has its own advantages and disadvantages, different smart assembly tools from different tooling manufacturers must be used. To allow a common interface and a common programming and analyzing software, the smart assembly tool architecture shall ensure a standardized interface/protocol. This enables the company tasked with manufacturing or assembly to integrate all of the different smart assembly tools into one company-owned system.

Regarding standardized interfaces/protocols, OPC Unified Architecture (OPC UA) should be mentioned. This is a machine-to-machine communication protocol for industrial automation developed by the OPC Foundation. The dedicated characteristics of this communication protocol focus on communication with industrial equipment and systems for data collection and control.

On the Fly

In the aerospace industry, products and parts are very complex. In the automotive industry, the products and parts are less complex, but the production rate is at much higher level. High transparency in the production flow is key for a healthy supply chain, good planning, and cost overview. It is therefore essential to present the progress of the assembly line at any point during production. Live monitoring and live data capture are important to support transparency and keep the quality level high (see the later section on Quality). Furthermore, live access to tools—at any time, from any location—could be very useful for manufacturing engineers and others to obtain and analyze tool status.

Connectivity

To enable storage of data and live access to all smart assembly tools, every single tool shall be linked to the Enterprise Resource Planning (ERP) system or to the company Manufacturing Execution System (MES). Edge and cloud services are also needed to store all data (big data).

Recommendations

Regarding ICT, smart assembly tools should enable and support industrial digitalization. It is recommended to use only battery-electric tools (such as those powered by a lithium-ion battery or another new battery technology). Quick and secure wireless connections should be present on all smart tools, such as Wi-Fi (2.4GHz and/or 5GHz) or another quick and secure wireless connection (e.g., 4G or 5G). Tooling architecture must be based on a standardized interface/protocol to allow easy integration in any company-owned IT infrastructure. Live monitoring and live capturing of data are necessary to interact with the tools to check and analyze their status and results. Furthermore, every smart assembly tool shall be linked to an ERP or MES, which could then be linked to an edge or cloud service for big data storage.

Cost Reduction

Energy Consumption

Cost reduction is one goal shared by every organization in every industry. Cost efficiency and competitiveness are required for long-term survival. In manufacturing and assembly companies, this could be achieved through several means, such as lower energy costs. (For example, pneumatic-driven tools have the highest energy cost—up to one hundred times more expensive than electricity.)

In the case of energy savings, smart tools can be programmed to shut off automatically when not in use and assembly cycles could be optimized with tool battery usage in mind. An electric, smart assembly tool could reduce the level of energy consumption by customization and optimization of the assembly process, for example the drilling process (e.g., when drilling through different materials with different speed and feed parameters).

Modular Assembly Tools

Besides reducing energy costs, it could be useful to create a modular smart assembly tool where the motor, sockets, different heads, and so on could be easily changed (e.g., with a quick change adapter). An auto-detection and/or auto-recognition system could be used for correct program selection (i.e., through a radiofrequency identification application). Currently, up to 1,000 task-specific tools are used in a modern aircraft assembly facility. Modular tooling ultimately results in less tooling, less tools in the production line leads to lower investment, and therefore reduced costs.

Integrated Quality Checks

Quality checks must occur on every fastener after installation. These post checks are manual operations today, but they could easily be completed with a smart tool (e.g., using sensors, software algorithms, etc.). This is leads to other cost- and time-saving elements. Additionally, detection of pass or fail (known as "OK/NO OK") assembly processes must be displayed to the operator immediately. This aspect is linked to the analyzing feature, which is described later in this report.

Limit Tools

Due to a smart assembly tool's ability to store and access multiple programs, the number of different tools can be reduced, which reduces cost overall.

Besides reducing the amount of tools, multiprogram smart assembly tools can reduce the time an operator spends searching for the right tool. This saved time can cut costs as well.

Recommendations

Smart assembly tools can help reduce manufacturing and assembly costs and ultimately lower the cost of the finished product. Due to cost-effectiveness, electricity should be selected over pneumatic power for manufacturing tools. Modularity and multiprogramming can also reduce the number of tools needed on the assembly line, which decreases investment costs and time spent searching for the right tool.

Process Control

Operator Guidance

Smart assembly tools should support the operator during the assembly operation. For example, during the tightening process, several parameters must be selected in

FIGURE 6. **A worker on an assembly line uses an HMI on a tablet computer to obtain data on the tool and assembly process.**

PopTika/Shutterstock.com

the right order (e.g., first torque, final torque, re-torque). Smart tools could also reduce human error by displaying this information to the user. This could be information like the next assembly steps or the current status about the process to the operator. One way of delivering this information may be human-machine interface (HMI) on the tool itself. Alternatively, smartphones, smart glasses, or tablets could be synced to the tools and serve as an HMI (Figure 6).

Transparency

To get more visibility in the future factory, smart assembly tools will need to ensure transparency of the assembly process at every point in that process. Progress of the work as well as wrong processes shall be visible at any time.

To that end, smart tools could provide valuable information regarding the

- Production site
- Working station and position
- Production line
- Project or product
- Serial number (as a Manufacturing Serial Number (MSN))
- Number of open and closed assembly processes
- Operator name (or even badge number)

For the future Industry 4.0 factory, it would also be helpful to link smart assembly tools with other current and previous assembly or manufacturing steps (i.e., smart gauging). Detection of nominal and off-nominal assembly processes should be displayed immediately to the operator by an OK/NO OK signal.

Data Collection

Every single assembly process should be captured by a smart tool and stored to enable

- Data archiving in case of product liability
- Data analysis for preventive actions, such as individual maintenance intervals or blocking a tool when it might proceed with a defect process
- Definition of the correct assembly parameters and optimal tool programming

Programming

Smart assembly tools should be able to be programmed with the right parameters, such as a tightening tool that can be loaded with the correct fastening regime (e.g., torque, angle, material type receiving the fastener).

Additionally, engineering and/or manufacturing engineering requirements may change over the course of a given production. Therefore, a new tool program would be needed. For example, torque values may change for a threaded fastener by the design office. In case of over-torque on a particular aircraft component, the stress office has to investigate this and a repair (fastener change) will likely be the obvious follow-up.

Recommendations

Smart assembly tools should be equipped with an HMI or be linked to a portable system, like tablet or smartphone, to guide the operator through the (changing) processes. An OK/NO OK signal should be clearly visible on the tool as well as a link to production site, the working station or position, the production line, the project or product, the serial number (as an MSN), the number of open and closed assembly processes, and operator identity. Data collection is necessary to enable long-term storage of data and analysis of activities. Due to electronic architecture, smart assembly tool shall be reprogrammable to cope with changing products and processes.

Quality

Analyze Data

As mentioned previously, machine learning and big data analysis are an integral part of Industry 4.0—and consequently smart tool solutions. Through the use of artificial intelligence (AI) and algorithms, manufacturing companies can detect any abnormal trend in the results or the behavior of the smart assembly tools. An "intelligent" tool should be able to track and analyze the results, and that intelligence can be stored directly on the tool or on a connected system (i.e., cloud service). This type of system is part of Statistical Process Control (SPC). Through SPC all information can be linked to all relevant tools. For example, if a bad batch of components is in production, the data from several tools can be analyzed to identify and confirm the quality of that batch, and whether the issue can be tracked to a particular fastener, lubricant, cutting tool, material, coating, and the like. If the batch is confirmed to be wrong, all tools can be equipped with a lockout to prevent further work on the off-nominal batch.

Optimized Calibration and Maintenance

Smart assembly tools need to be calibrated, maintained, and even repaired like any other assembly tool or system. Additionally, tighter tolerances from the product design office may require better accuracy from current tools and the processes. This can be achieved with optimized calibration and maintenance intervals enabled by machine learning and predictive maintenance features (Figure 7). By analyzing captured tool data, trends can be detected and interpreted as

FIGURE 7. **Calibration of a manufacturing tool.**

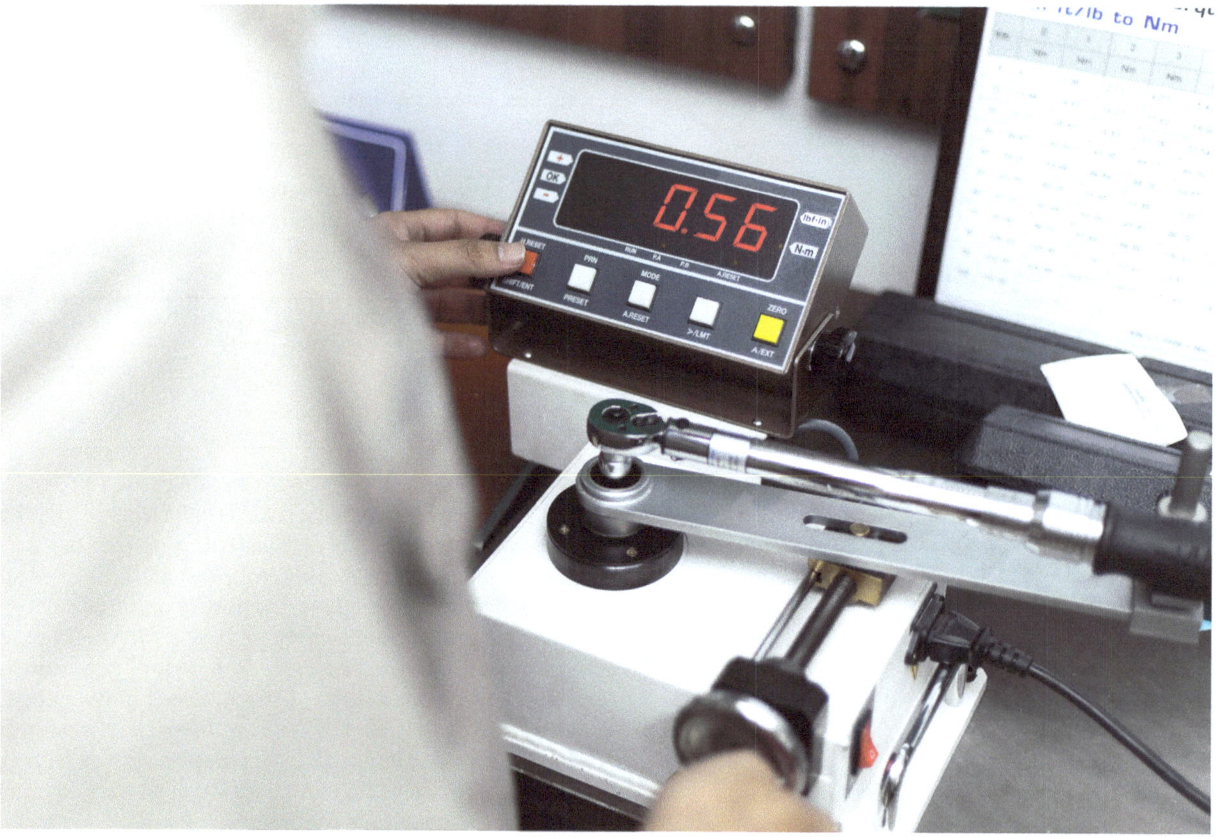

FarmChung/Shutterstock.com

FIGURE 8. **Geo-localized devices.**

abnormal. In addition, it can be locked if the tool recognizes an abnormal trend by analyzing the results and tool status as mentioned previously.

Geo-localization

As shown in Figure 8, all various smart assembly tools can be connected to the local ERP or MES. In addition to the analyzing features, a smart assembly tool should tell to the system "where" it is working—such as which production line, station, production part—but also the actual geographic location of the production part. This can also allow the tool to be locked out whenever it is outside of the dedicated usage area, a feature especially useful for preventing certain tools from leaving certain areas—as well as deterring theft. Another example would be geo-localized lockout when an operator wants to install a fastener in a wrong hole/position.

Preventive Warning Systems

All the abovementioned analysis and detection features can be used to enable a preventive warning system. A system like this could detect abnormal trends, misuse of the tool, incorrect program selection, and so forth; it can be used to avoid mistakes before they occur. Ensuring complete traceability can help to get the right level of predictive maintenance, before the tool needs a repair or is out of tolerance. Alternatively, this traceability can be used as a reason to simply lockout a tool that requires maintenance (and would otherwise negatively affect product quality).

Certification

Real-time detection of defects, mistakes, misuses, or any other kind of quality aspect mentioned before is vital to maintaining appropriate levels of quality assurance. However, beyond helping to maintain this quality assurance, smart assembly tools can also be used to certify the operation or assembly step (Figure 9). This can be seen as the highest level of smart tool functionality. To enable this feature, all levels mentioned in Figure 9 must be in place to allow an overall assessment of the performed assembly task.

Recommendations

Throughout this report, the advantages of implementing smart assembly tools are discussed in terms of quality. Analyzing captured tool data can contribute to optimized and individual calibration and preventive maintenance intervals. In addition to this, geo-localization can be leveraged to assign the results and data to specific products or processes. With these features, a preventive warning system and certifying procedures can be established to avoid quality issues in advance and achieve transparent quality.

FIGURE 9. **Smart Assembly Tool Technology Levels linked with their unsettled industrial domains.**

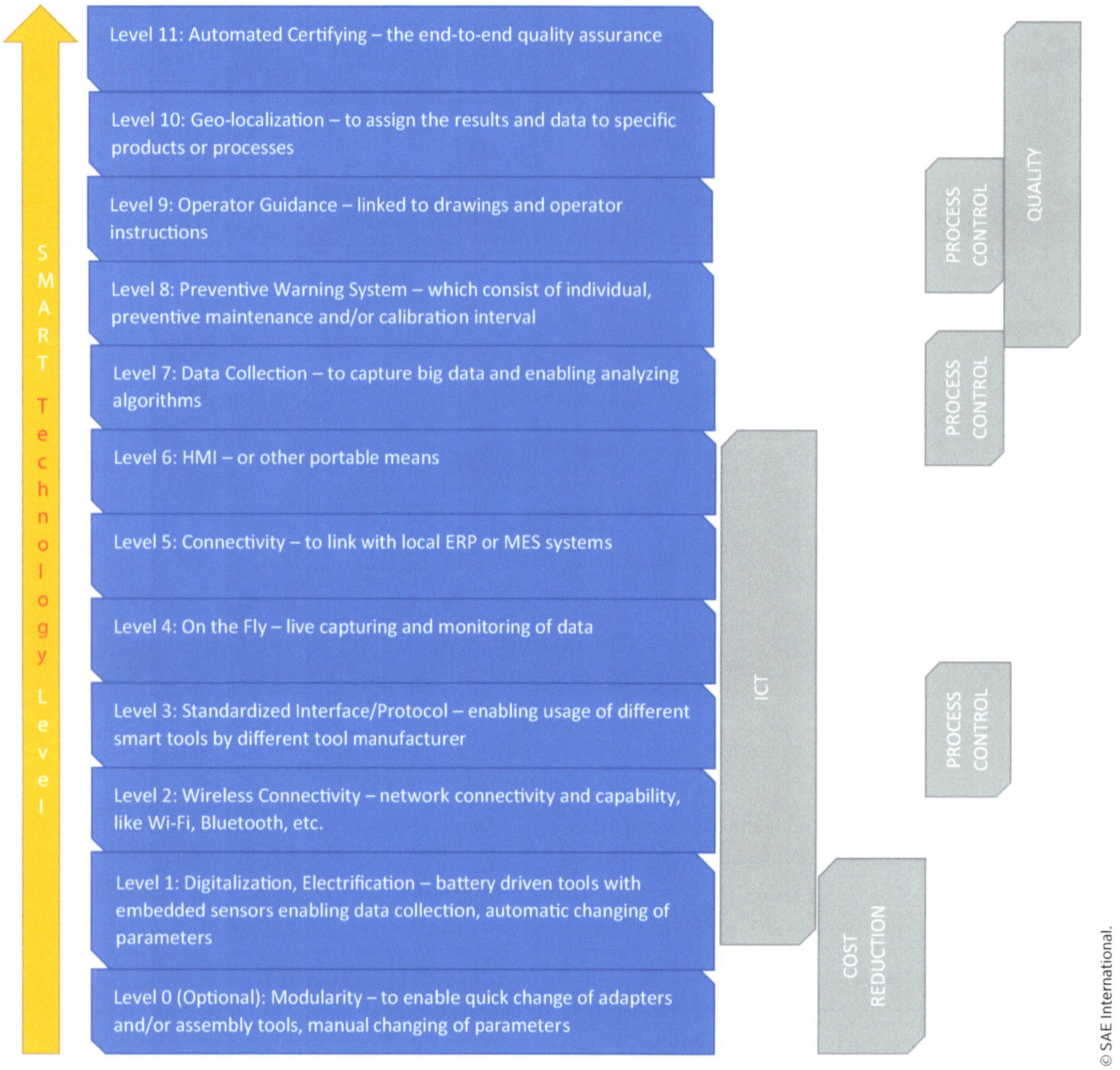

Summary

In this SAE EDGE Research Report, the different features of smart assembly tools are discussed and key unsettled definitions of the roles in the Industry 4.0 environment are mentioned. Smart assembly tools are used in all industries—including automotive, aerospace, and space—for measurement, inspection, gauging, drilling, and installation of all existing fastening systems (e.g., lock bolting, riveting, blind bolting, and tightening operations). Smart and intelligent assembly tools can enable a range of new functions inside the Industry 4.0 environment, and they offer tremendous potential value for several industrial concerns, such as quality assurance, process control, cost reduction, and data analysis.

The following definitions can serve as a kind of standard to sum up all features of smart assembly tools and to sort them by technical level:

- **Level 0 (Optional):** Modularity, to enable quick change of adapters and/or assembly tools, manual changing of parameters
- **Level 1:** Digitalization and electrification to enable a battery-driven tool with embedded sensors for data collection, automatic changing of parameters
- **Level 2:** Wireless connectivity and network capability, like Wi-Fi or Bluetooth, for connecting to a monitoring system

- **Level 3:** Standardized interface/protocol to enable usage of different smart tools by different tool manufacturers
- **Level 4:** Live capture and monitoring of data
- **Level 5:** Connectivity with local ERP or MES systems (integration at customer side/ecosystem)
- **Level 6:** Operator guidance via HMI or other portable means
- **Level 7:** Data collection to capture big data and enabling analyzing algorithms
- **Level 8:** Preventive warning system, which consist of individual, preventive maintenance, and/or calibration intervals
- **Level 9:** Operator guidance (linked to drawings and operator instructions)
- **Level 10:** Geo-localization to assign the results and data to specific products or processes
- **Level 11:** Automated certifying/quality stamp to enable 100% quality assurance

The technical levels according to the four unsettled domains are summarized in Figure 9.

These four technical fields identified in this report (i.e., ICT, cost reduction, process control, and quality) need to be discussed in greater depths with aims of, first, clarifying the scope of the industry-wide alignment needed, and second, prioritizing the issues requiring standardization.

SAE EDGE Research Reports

SAE EDGE Research Reports, like the present report on "Unsettled Technology Domains in Industrial Smart Assembly Tools Supporting Industry 4.0," are intended to push further out into still unsettled areas of technology of interest to the mobility industry. SAE launches these reports before attempting to form a joint working group, let alone a cooperative research program or a standards committee.

SAE EDGE Research Reports are intended to be quick, concise overviews of major unsettled areas where vital new technologies are emerging. An unsettled area is characterized more by confusion and controversy than established order. Early practitioners must confront an absence of agreement. Their challenge is often not to seize the high ground but to find common ground. These scouting reports from the frontiers of investigation are intended merely to begin the process of sorting through critical issues, contributing to a better understanding of key problems, and providing helpful suggestions about possible next steps and avenues of investigation.

SAE EDGE Research Reports, therefore, are fundamentally distinct from the more formal working groups approach and far removed from the more mature research program and standard's development process.

Next Steps for Smart Assembly Tools

This publication should be considered only as a first step toward clarifying the issues around the definition of smart assembly tools for Industry 4.0. The intention behind this and other SAE EDGE Research Reports is to start a dialogue among interested parties on important industry-wide topics that require further attention. The expectation is that these explorations of unsettled areas of technology will lead to the formation of working groups and, ultimately, committees that can address and resolve the issues they raise, producing a framework for developing a common vocabulary of definitions, best practices, protocols, and standards needed to support continued progress toward safer and more innovative automotive products.

The experts' collaboration that gave rise to this publication demonstrated a great willingness on the part of the industry to define the terminology, procedures, and eventually the standards needed to define smart tool technology to move ahead as quickly and efficiently as possible.

This SAE EDGE Research Report on smart assembly tools identifies the following key topics for further pursuit, both through continued informal discussions among industry practitioners and through more formal working groups:

- Clarification and alignment on the ICT infrastructure, common understanding of different industrial backgrounds
- The cost reduction required for different industries
- The issue of process control for assembly tools and technology
- Manufacturing and assembly quality in the Industry 4.0 era
- Establishing of the scope of the industry-wide alignment and prioritizing the issues requiring standardizations of smart technology levels for assembly tools

Recommendations

The overall recommendations of this SAE EDGE Research Report can be summarized as follows:

1. Identify requirements and definitions defined by the technical fields identified in this report (i.e., ICT, cost reduction, process control, and quality).
2. Arrange a hierarchy for the four technical fields to identify different levels of smart technologies.

3. Create standards for different levels of smart assembly tools.
4. Define a roadmap for standardizing smart technology levels for assembly tools. Figure 10 proposes an initial guideline to support manufacturing engineers through this process.

Definitions

AI - Artificial Intelligence

ERP - Enterprise Resource Planning

HMI - Human-Machine Interface

Hz - Hertz

ICT - Information and Communication Technology

IT - Information Technology

MES - Manufacturing Execution System

MHz - Megahertz

MSN - Manufacturing Serial Number

SPC - Statistical Process Control

Wi-Fi - Synonym for Wireless LAN, or "WLAN"

WPA - Wi-Fi Protected Access

Acknowledgments

Recognition should go first to all the participants, many of who also provided feedback on the draft version of this publication. Without their input and initiative, this SAE EDGE Research Report would not have been possible.

Lars Burkhardt, Atlas Copco
Baptiste Bertrand, Airbus Operations SAS
Pierre Deffrennes, Desoutter Tools
Florian Hanke, HS-Technik
Thomas Schönbeck, Airbus Operations GmbH
Sebastian Ebeling, Siemens Gamesa

The author of this document together with the SAE Team responsible for its creation join in expressing our deepest appreciation to all the individuals mentioned above.

Thorsten Roye, Dipl.-Ing., M.Sc.
Airbus Operations GmbH

References

1. Holleman, W., Whinnem, E., and Wrede, M., "Smart Portable Tools: The Practical Solution to Agile Assembly," *SAE Int. J. Aerosp.* 6(1):180-187, 2013, https://doi.org/10.4271/2013-01-2295, accessed July 29, 2020.
2. Desoutter 4.0, Accelerate you transformation, Desoutter, https://www.desouttertools.com/uploads/documents/5b9911664a64b_DESOUTTER_4.0_Book.pdf, accessed: July 29, 2020
3. Smart Connected Assembly - Powered by Data, Atlas Copco, 2019, https://www.atlascopco.com/de-de/itba/local/smart-connected-assembly-roadshow, accessed: July 29, 2020

Contact Information

EDGEresearch@sae.org.

www.ingramcontent.com/pod-product-compliance
Lightning Source LLC
LaVergne TN
LVHW070942210926
848482LV00003B/282
* 9 7 8 1 4 6 8 6 0 2 4 2 5 *